# Contents


Preface

How the Oilfield is Organised

The Drilling Rig

Before Travelling Offshore

Accommodation Block

Sharing a Cabin

Derrick and Drill Floor Operations

Safety

Work Permits & Safety Reporting Cards

Severe Weather Conditions

The Working Environment

Technology

The Ocean

New Recruits

Conclusion

Figures

## Preface

The aim of this book is to give a well-rounded and light-hearted view of the oil and gas industry and of the activities of a petroleum worker while searching for hydrocarbons beneath the surface of the planet. My hope is that it will inform and inspire the younger generations - the future leaders of the industry, others who may be considering a career change into this area or those who work or have worked in the industry to reminisce about the time they spent here. Offshore oil and gas extraction has been well practiced since 1896 and has been among the most complex and forward-looking engineering industries on earth. The energy produced from the "black gold" far underground is still used to power a large part of human activity today.

Due to its relatively recent discovery, the lives of those who have worked regularly offshore has been well documented. Therefore, it has been possible to analyse, in acute detail, various aspects of normal offshore life that vary greatly from the lives of those who work on land.

I was born into an oil family myself, with my mother, father, an aunt, two uncles and grandfather all spending their careers supporting the industry. My own career began here as well, with many years spent on a great number of offshore installations around the world.

As well as my own experiences and knowledge, various colleagues, relatives and many others have contributed content and shared their experiences to ensure that the text within was made as complete and accurate as possible.

*Figure 1 - Offshore Installation* [1]

Throughout this short book I have embarked to cover the life of an offshore worker in enough detail for readers to come to their own conclusions should they desire to pursue a career here and, more generally, regarding the image of the oil and gas industry to the public. Hydrocarbon companies have been hugely successful and some remain among the cornerstone indicators of entire country economies today. It has, however, regularly been singled out for controversy and cause for protest; this industry does not agree with everyone. It is dangerous, it is exhilarating, it has committed mistakes yet it is a unique career experience.

There can be no doubt that oil and gas extraction remains to be one of the world's most important industries. It has accelerated the development of the modern world; fueled almost every mode of engine transport, enabled heat and electricity generation, asphalt and the development of various plastics. It remains a very lucrative, challenging and rewarding career with opportunities around the globe. Even once eventually replaced by other forms of energy, it will be remembered as a pioneering leader for improving global standards of living and it has been prevalent in the lives, in some way, of every person on earth.

## How the Oilfield is Organised

The oil and gas industry is massively complex and diverse. A typical oil field life cycle spans an initial expedition to scan what might lie beneath the surface of the earth, moving an oil rig to the location, drilling, producing and maturing, right through to the moment the last pieces of the oil rig return to land following disassembly. Furthermore, the hydrocarbon journey involves being brought to the surface from thousands of metres underground, being transported onshore, refined, sold and finally supplied to the market as a customer refuels their car from a petrol station. There are so many wide-ranging activities and challenges that it is simply too diverse for one company to carry all the in-house expertise necessary to cover all operations. Therefore, multiple companies exist to support various aspects of the industry and these can be split broadly into three groups.

The Oil Companies occupy one of these groups. An oil company generally reaches a buy or lease agreement with a local government to explore an area of the ocean. Using data retrieved from initial seismic surveys of the area, expert teams analyse and weigh up the probability that hydrocarbons are present, and whether it will be profitable to extract, produce and sell. Once all this is considered and approved, the oil company arranges for an oil rig to station itself in the area and to drill a well. If hydrocarbons are discovered, the oil company endeavours to develop the field and produce as much as possible and transport it, through various means, to a refining plant. Here oil and gas is refined and divided into its constituent forms for appropriate use (petrol, lubricants, plastics etc.) and then uses its marketing teams to facilitate the distribution and sale of these products to the world.

Service companies also play an important role in the industry. These companies perform tasks that require specific tools, equipment and specialist knowledge. For example, the offshore industry cannot operate without helicopter and supply boat service companies to transport people and equipment between the installation and company workshops. The oil company does not build or maintain these themselves and does not have the capacity to do so. As a result, service companies specialise specifically in these operations. Furthermore, on the oil rig, many operations are supported by service companies. Safety aspects, such as life boat maintenance, life jacket inspections and general rig maintenance are passed to service companies to be taken care of.

Various stages of hydrocarbon discovery, drilling and production are also covered by a service company. Seismic surveys from these firms aid the oil company to decide whether to commit to exploring the area for hydrocarbons. The service company provides the drill bits used to drill a well, the provision and treatment of mud to lubricate the drill bit while drilling and to stop the well from collapsing; and the wireline logging tools used to identify more specific well properties. Furthermore, the casing design used to make the well secure against blow out and collapse, mechanisms for transporting hydrocarbons from the formation to the surface and then along pipe lines to the refineries are also outsourced to service companies.

The drilling contractors are the third group, which design and construct oil rigs before providing the crew to execute day-to-day operations while at sea. Oil rigs are impressively complex, serving all the required functions to drill for, retrieve and process oil and gas energy thousands of metres below the surface while remaining a safe and reasonably comfortable place of work for those who find themselves marooned there for some stretch of time. To ensure the smooth daily running of the oil rig and to support the oil company's discovery efforts, a rig crew is also provided. They cover operations for the entire duration of the project – twenty-four hours per day, seven days per week and three hundred and sixty-five days per year. The crew includes a team of drillers and roughnecks to execute the main task of a drilling contractor: to drill holes in the earth. Furthermore, cooking and cleaning staff, the medic, electricians, mechanics, an overall installation manager and other essential personnel are provided to make the entire operation possible.

When an oil company decides to explore a potential hydrocarbon site, it sends out a proposal and commences a bidding process with prospective drilling contractors and service companies. If these companies desire to play a part in the project, they submit their bids to the oil company for review. Once all bids are considered, companies are selected to provide specific products and services throughout the project. Agreements between the oil company and the drilling contractors and service companies vary in several ways. Payment for services can be based on many factors and may include: number of feet drilled, contractor service delivery efficiency, evidence and demonstration of prospective company's commitment to health and safety or based on project completion efficiency. Work may also be awarded for a single well, a single campaign or may be awarded for all the company's global operations for a set time period.

The fluid and respectful collaboration of these three sub-groups is critical to ensure that the industry remains safe, efficient and profitable while remaining a desirable career choice for many and an attractive investment for shareholders.

## The Drilling Rig

There exist a great number of drilling rigs both on land and in the world's oceans. Here we will concentrate on the ocean, where roughly one third of the world's oil has been extracted. The oil rigs (also known as installations) are generally split into 4 variations: Platforms, Jack-ups, Semi-submersibles and Drill ships. Other types do exist although in far fewer numbers than the four mentioned here.

*Figure 3 - Oil Platform* [iii]

Platforms were the first oil rigs to be built following the discovery of hydrocarbons in the ocean. The platform sits on steel or concrete legs that extend from roughly 30m above the ocean (to avoid damage from the biggest waves) to the sea bed so that the rig is fixed into position. 10's of oil and gas wells can be drilled from the platform and bring hydrocarbons to the surface for many years

Jack-ups are the second generation of rig. They also have legs constructed from steel that extend to the ocean floor before motors "jack-up" the rig deck to remain above the waves. The depth of water they can operate in is dictated by the length of the legs, so they tend to remain in shallow water. The great advantage of a jack-up is that, once a well has been drilled and secured, the legs can be driven up from the sea bed and the oil rig can be towed to other locations. Therefore, a jack-up can take charge of, and travel between, a series of wells spread over a very large area.

Figure 5 - Semisubmersible Oil Rig [liv]

More recently - the semi-submersible. Without legs extending to the sea bed, the semi-sub is more mobile than the jack-up and is not so dependent on sea depth for its operations. Large ballasted pontoons just below the surface of the ocean provide rig buoyancy, and steel legs support all decks and equipment above the waves. Hence these are generally used for deeper water exploration. Wells can be drilled and if nothing is found then the installation is towed by tug boats to the next potential area for the subsequent operation.

*Figure 6 – Drillship* [v]

Drill ships are the latest, most manoeuvrable edition. With all the capabilities of the semi-sub, some have the ability to operate in water depths greater than 3000m. Without the requirement for other vessels to transport it between locations, the drillship can sail itself between potential sites for its operations.

Once personnel arrive on board, the oil rig is effectively their home and it consists of various sections. A large circular helideck allows the helicopter to land after transporting workers between the oil rig and the shore. The accommodation block sits directly below the helideck. All bedrooms, kitchen, galley, offices, exercise facilities, recreation rooms and medical facilities are located here.

Next to the accommodation are the cranes and decks. Cranes work with the supply vessels, that sail back and forth to the oil rig from the port with equipment, to lift whatever is required onto the deck to carry out operations. Various offices and workshops for the on-shift personnel are stationed below the top deck, with examples including: welding room, mechanics' room, electricians' room, remotely operated vehicle (ROV) point and power generators.

Furthest from the accommodation block is the derrick and drill floor. The derrick is the instantly recognisable steel tower and is generally the tallest structure on the entire installation. Underneath this is the drill floor where the drilling activities take place. Furthermore, immediately surrounding the derrick is all the equipment required immediately for day-to-day operations and provides space for safety critical and operationally critical tools. Therefore, if needed, they can be located and utilised at short notice without support from the cranes.

## Before Travelling Offshore

The offshore oil rig is a destination accessible only by helicopter or boat. The helicopter is much faster, hence is widely exploited as the preferred mode of people transport. Every individual that travels offshore on a helicopter must pass some form of minimum industry standard training (MIST) and complete some practical scenarios. Beginning with classroom based training, prospective offshore workers are introduced to the oilfield, safety measures, an insight into what to expect and, most importantly, how to jettison in an emergency escape and survival situation.

Following classroom based training, the class practise various escape and survival techniques should they enter the ocean. In designated pools, teams practise treading water in groups when stranded, setting up life rafts and entering them, escaping via each of the available mechanisms (escape ladders, ropes, pulley systems, escape chutes) and finally familiarise with techniques to exit a helicopter during a landing on water. Helicopter escape training involves being lowered from height into the water, using the breathing system attached to one's life vest, punching out the side window and exiting the aircraft. The helicopter may remain above the water, become submerged or even flip upside down underwater. All individuals must prove they are capable of escape in each scenario.

*Figure 8 - Helicopter Escape Training* [vi]

Further to the MIST training, an independent medical assessment must also be completed prior to travel. Medical support offshore is available but is of basic standard. Any complications or advancing illnesses severely reduces the chance of survival as the nearest hospital requires a helicopter journey to reach. A person's size and weight are also considered and, critically, the individual's shoulder width must be verified to ensure they are able to escape from the aircraft through the window if required.

As mentioned earlier the helicopter, also known as the "chopper", is the mode of transport. After landing on the helideck, assigned members of the oil rig (known as the deck crew) meet the helicopter, open the door and personnel exit the helideck via a set of stairs. Often a single chopper visits more than 1 rig during a flight, carrying passengers assigned to each rig. Therefore, it is critical to stay alert and not exit the helicopter on the wrong oil rig! This has been known to happen, although the victim will never admit it to their peers. As a result, the oil rig's name is displayed in large lettering on the helideck and further serves as

an indicator for the pilot that they have reached the correct destination.

*Figure 9 - Survival Suit* [viii]

In areas where a potential descent into the ocean would subject the passengers to cold sea temperatures, a diving dry suit called the flight suit is used. At the heliport, prior to ascending the helicopter, a flight safety video reminds the passengers how to put on their flight suit and covers basic flight safety rules and protocols should a landing on sea be necessary. The safety video is of paramount importance; I recall one occasion during a flight safety video, passengers were engaged in conversation and not reviewing the film as it was being played. This behaviour was reported, the individuals were removed from the flight and other company representatives had to be mobilised at short notice. This caused great embarrassment to the individuals involved and to their employing company.

When a flight suit is necessary, most get into it at the last possible minute! Its neck and wrist seals can make the suit hot and uncomfortable to wear; although, if called upon, it can save one's life. It also houses breathing apparatus which will allow the user perhaps 30 seconds of air to push open one of the side windows and exit the helicopter should they become submerged under water during an ocean landing. Helicopters can make multiple flights per day, each with two pilots and up to 16 passengers.

All commercial airlines have strict baggage limits; similar restrictions are apparent in helicopter flight. Generally, 1 piece of luggage of 12Kg is permitted for offshore travel. Sorting out 12Kg for all clothes, toiletries and personal protective equipment for a three-week trip is challenging. You may bring certain brands of mobile phone (depending on battery reliability), books and razors. However, among other banned items, no sharp objects, drugs, alcohol or toxic chemicals may be transported offshore in personal luggage. Medical prescriptions are permitted but they must be declared and passed to the offshore medic once on board so that, if anything were to happen to the individual while isolated and many miles from land, the on-board crew would have an idea of how to support one's illness.

## Accommodation Block

The accommodation block is as close to a home from home as it gets for workers. Drilling contractors exert a huge effort to design entire rigs but the accommodation block receives unparalleled attention to be as comfortable and efficiently designed as possible. This provides the best environment for an interactive and productive workforce.

One of the main communal areas is the galley and is often the favourite location for the workers to relax after working outside in all weathers. Normally open between 0500-0700, 1100-1300, 1700-1900 and 2300-0100 (breakfast, lunch and dinner for both the day and night shift crews), offshore catering staff are employed to provide a range of healthy and energy rich meals for the staff onboard; the food is delicious! Meal sizes and options were scrutinised following a survey of the average weight of an offshore worker over time, where a continual increase was noticed in the 30 years up to 2014. Since then the industry has taken on initiatives to promote a healthy lifestyle, reverse this trend and to enhance workforce productivity. However, from personal experience, I have yet to experience an offshore installation with a poor range of food options and a tightened attitude to portion sizes. There is always an array of unhealthy desserts available!

As the canteen generally prepares healthy meals, most installations keep a tuck shop on board as well. This normally opens for 15-30 minutes per day, supplying individuals with sweets, fizzy drinks, cigarettes and basic toiletries if required. These cost money so it is important to remember to bring some cash in case you run out if supplies.

Recreation rooms are the second most used communal area and are often used as the pre-shift meeting room for the employees prior to starting work. Outside these times it is a hub for viewing television and films. Occasionally I have found free-for-use iPads in these rooms and charging points. As well as this, it has become popular as the post-meal nap room prior to returning to work for the second half of the shift. Furthermore, if one is lucky, a music room may exist as well with a selection of a set of drums, guitars, electric piano and other instruments that are free to use during leisure time.

The medics room is also in the accommodation block and always has a secured cupboard with all the prescriptions for individual workers. I have often found that medics find themselves issued with new "gadgets" and are always on the lookout for a passer-by to test them on. For good reason as well; on one trip offshore, a new random drug testing machine had been installed and the person randomly tested carried illegal drugs in their system. They were promptly escorted from the oil rig on the next helicopter. A person working under the influence of drugs, alcohol etc. is a significant risk to the other workers and to the entire operation offshore due to the complexity and hazards associated with the equipment used. As a result, all oil and gas companies have a zero-tolerance policy on drugs and alcohol in the workplace.

The gymnasium is the only location to exercise. I have heard stories that previously, workers used to run in a circle around the helideck in their off-shift time. However, the risk of an individual using this area when a helicopter is forced to make an emergency landing, or of falling over the edge into the sea, means that this is no longer practised in most areas. Gym machine options vary depending on available space. Some are fantastic; with endless dumbbell racks and running machines, a sauna and, in some cases, I have been lucky enough to visit platforms with a squash and volleyball court. Most, however, are not as luxurious as this and one may find themselves merely doing press-ups in their room or using a skipping rope in the corridor with others passing by.

Offshore life was once very isolated, with limited means of communicating with colleagues onshore in the event of challenges and support to receive more equipment. It was also difficult for workers' families when they disappeared for weeks at a time. More recently, phone booths and wi-fi make it much easier to maintain relations on-shore for both business and personal communication. Some installations are fibre optics hubs, enabling video communication. This was hugely beneficial for me when communicating with my partner when we lived on different continents and video calls with the family one Christmas day.

**Sharing a Cabin**

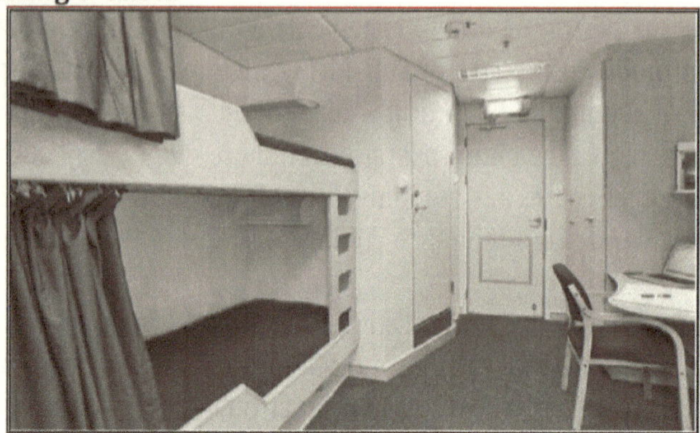

*Figure 11 - Oil and Gas Installation Cabin* [viii]

Sleeping offshore varies from onshore in several ways. Firstly, the room is shared and is yours for 12 hours only. Most rooms contain two beds; one for the day shift worker and one for the night shift worker. In earlier editions of offshore rigs, four or eight-man cabins were standard. Once one is "on shift", they cannot return to the room they slept in as it is now for another person to use. Be sure to lock all your belongings away and do not leave anything lying in the room prior to the end of your 12 hours, or else! Strangers are partial to using other's belongings, especially body wash, deodorant and shaving accessories!

Offshore bedrooms are very comfortable. Bed sheets and towels are changed every day by the cleaning staff on board and laundry can be picked up and delivered to your bedroom door as well. The room is warm (or is air conditioned in climates closer to the equator), with good showering facilities and most have a television, desk and telephone as well. The room is dark to allow the user good quality rest no matter which shift they are working during their time offshore.

All bedrooms have a survival bag for each occupant to store their offshore flight suit. It also contains: a life jacket, torch, fluorescent glow stick, gas mask and heat resistant gloves (if you must escape and have no option but to navigate through burning or charred sections of the installation to safety). Do not drink the water directly from the tap! Some of the installations have been around for decades without refurbishment so may be contaminated!

There may be occasions where two individuals share both the same shift and the same room. This is all part of the experience and, as long as both occupants are considerate, it works smoothly and can even result in new friendships!

## Derrick and Drill Floor Operations

The drill floor and derrick is the location where the well is drilled. Operations mainly involve different sizes and types of drill bit on the end of drill pipe to descend underground, making holes and adding joints of drill pipe as the hole increases in depth. It also involves running various sizes of steel casing to stop the well from collapsing in on itself and to set cement. Hardened treated cement between the sub-surface formation and the outer side of the casing create a protective seal against a potential hydrocarbon escape from underground and blow out of the well at surface.

*Figure 12 - A drill crew on the dill floor [ix]*

Highly skilled drill crews control and supervise all operations on the drill floor. If not drilling, they support service companies during various stages of well drilling and development to setup, execute and disassemble (also known as "rigging down") their respective operations. This includes drilling and completing the well, determining hydrocarbon properties and attaching testing and well production facilities to process hydrocarbons as they are brought to the surface. Drill floor operations are somewhat automated from a drillers shack in one corner of the drill floor. However, a large amount of manual, energy intensive labour is involved and it is important that the time is taken to assess the risks and then perform each operation as safely as possible.

## Safety

Safety is of paramount importance throughout the industry as the work undertaken can be dangerous if the time is not taken to consider and eliminate or reduce risk wherever possible. Almost every company working in the industry has a top priority of ensuring safety for those that work with and support their operational ambitions. On the oil rig itself, there are too many systems and controls to describe here to manage the various daily activities. The important safety features for all employees, however, are the ones that will allow escape should a fire or an abandon rig scenario arise. Disaster situations and escape mechanisms are illustrated to workers as soon as they arrive on board, either through practical demonstration and practise or by use of a video describing the safety features of the installation.

The life boats provide the primary and safest method of escape. Life boats are situated all around the rig and have a combined number of seats far exceeding the maximum capacity of the rig at any one time. The greatest proportion are stationed by the accommodation block, where most workers stay most of the time. This will allow all personnel to vacate even if some of the life boats are unreachable at the time of abandonment. Personnel are assigned a life boat once they arrive on board and are required to line up, put on life jackets (which have a light and a whistle for attracting attention) and record their names to confirm everyone is present. After entering the life boat, a trained pilot controls a winch to commence a controlled descent into the ocean and then operates the engine to steer away from the installation to safety.

Should one find themselves alone and with all the lifeboats already jettisoned, there are also escape ladders and hanging ropes scattered around the rig with life jackets close by. On descending one of the ladders or hanging ropes, individuals can reach small decks at the bottom of the legs and find themselves much closer to the ocean where, hopefully, they can be rescued or can swim to safety. These are alternate escape options; however, they are more dangerous due to the increased likelihood that the individual will enter the ocean and, if isolated, can quickly develop hypothermia.

Locations also use escape chutes. These are self-contained slide ladders that allow a person to descend towards an awaiting lifeboat and are only used in specific offshore sectors. Their advantage over the ladders and escape ropes is that there is no possibility for an individual to slip and fall off an escape chute, hence is guaranteed to make a safe, although possibly slower, descent.

Aside from evacuation mechanisms, the safety of all workers is demonstrated through stringent personal protective equipment (PPE) requirements. Every individual leaving the accommodation block may only do so if wearing full PPE. Clothing consists of: a hard hat, safety glasses, full length coveralls, gloves and steel toe capped boots. In areas of increased hazard, such as excessive noise, ear plugs or protectors are mandatory prior to entering. Hard hats do go out of date! Hence it is important to source a new one if the existing one is out of date or damaged. Glasses cleaning stations are situated around the rig to ensure clear vision while working. Coveralls are fire retardant and must be thrown away if ripped or damaged and steel toe capped boots are mandatory to protect feet in case of a dropped heavy load. Various types of gloves may be worn on board depending on the task to be performed.

Supply boats are also in close vicinity to the oil rig and, if called upon, can quickly move into a position and activate water hoses to alleviate fires on the installation. These boats are often shared between installations near each other and are on continuous standby.

## Work Permits & Safety Reporting Cards

All work is controlled by permits to ensure safe execution of daily operational tasks and to avoid hazards by work scopes interfering with each other (e.g. one permit involves welding and sparks and another, in the same area, involves the use of explosives). If at any point during an operation the scope of work changes and is no longer covered in the description of the work on the work permit, the operation must be stopped and a new work permit completed before continuing.

The best time to create and, most importantly, obtain signatures for a permit is during the pre-shift meeting when all personnel are present. Individuals embarking to initialise permits after the morning meeting face a painful search of the entire platform to find the individual and obtain their signature of approval. From first-hand experience, this usually involved multiple trips inside and outside the accommodation block, removing and putting on personal protective equipment and multiple phone calls, resulting in up to a 3-hour delay for starting a specific daily task. Unless given prior consideration, all permits are valid for a maximum of 12 hours to allow new permits to take any potential changes during the job into consideration.

As mentioned previously, personnel safety is the number one priority for organisations involved in the industry. Each have their own safety initiatives to keep all workers engaged, looking out for each other and contributing to the continual improvement of existing safety measures. On an offshore oil rig, this usually comes in the form of a small card where one writes down any substandard situation that they have encountered and the preventative or mitigative measures taken to reduce the risk of damage. Reports are collected and examined by the management team to monitor operational activities and act if, for example, an object or area has become unsafe or if an individual or group could perform an operation in a safer manner. Regular rewards and recognitions are given to those who take the initiative and action to intervene and stop a substandard condition from occurring or from getting worse. Often these individuals are recognised in front of their colleagues as well to promote good quality safety reporting. Furthermore, companies reward groups that pass certain milestones without injuring one of their group. Accomplishments such as "one hundred days without injury" or "one million-man hours without a lost time incident" usually result in the award of t-shirts, jackets or vouchers for the group that achieved such a feat.

Most recently, as the industry aims to achieve a zero-injury environment, most installations employ a 1-card-per-day minimum reporting standard to encourage active participation by all the workforce. This has sometimes led to a conflict of number of cards vs quality of reports passed to the management team. Despite this, there can be no doubt that the industry today is safer than it has ever been; yet there is still room for improvement and to eliminate all work-related injuries.

## Severe Weather Conditions

*Figure 14 – Severe weather offshore* [x]

Seasonal weather conditions are amplified out in the ocean, where there are no barriers to curb or reduce the effects of oncoming pressure fronts. The waves grow in excess of 30m in the most extreme, brushing the underside of the oil rig and rendering outside work impossible with too much of a risk to life. This is usually accompanied with hurricane force winds and sometimes no individual is permitted to leave the accommodation block. The medic is usually the busiest member of staff throughout, treating others for sea or motion sickness on the installations that function without steel or concrete legs fixed to the sea bed.

The rig cranes suffer the most due to offshore wind. Due to their height and exposure to the elements, most have a maximum wind speed operational rating, above which they cannot execute lifting operations. This can cause severe delays to operational activities and, as a result, weather is a major consideration for planning daily rig activities.

In tropical locations, spectacular thunder storms also hamper the progress of operations with spectacular light displays. During these events, it is all too easy to feel insignificant and very vulnerable to mother nature. Of course, any offshore operations that make use of explosives or the arming of explosives must be delayed until a thunder storm has passed by, due to the risk of stray currents.

Occasionally entire installations must be secured and then evacuated by all staff to avoid catastrophe. This has been practised in sectors such as the Gulf of Mexico, flying all workers back to the shore prior to a hurricane sweeping through.

*Figure 15 - Partially sunk rig following extreme weather* [xi]

From personal experience, if you ever find yourself stranded on an installation during exceptional weather, I advise you always bring a book and some exercise clothes to pass the time.

Aside from the oil rig; the supply boats and helicopters are also heavily reliant on the weather, with hazardous conditions essentially rendering these services useless and unable to support operations. As a result, advanced planning is always necessary ahead of challenging natural events. Helicopters are prone to fog, lightning, high winds, snow; any natural situation which reduces visibility or flight safety beyond a reasonable level and would also hinder any rescue team should the helicopter enter the water during a flight. Furthermore, supply boats cannot be operated in certain weather conditions, where the swell can be too rough to transfer equipment and containers to and from the vessel. High waves and bad weather reduce vessel control and it is hence prohibited from approaching an oil rig in case it comes into contact with the side or one of the legs, potentially causing damage and endangering the entire rig crew, as well as the vessel crew.

## The Working Environment

Working offshore is definitely more hazardous than working on land. However, if necessary precautions are taken, and they are taken on every installation I have visited, all risks can be managed effectively. Some of the installations have graced the ocean for almost 50 years. Clearly this will have taken its toll on the rig structure. If you reach the conclusion that this career appeals to you, please be aware of the following common hazards that one should always be aware of.

Asbestos was used for wall fireproofing in the first oil rigs during construction. Although it was largely removed once it's health hazards were exposed, some asbestos has remained in out-of-reach locations. These are clearly marked and should be avoided if possible. Asbestos related illness, resulting from inhalation, includes cancer, asbestosis and other fatal illnesses.

Hydrogen Sulfide is a poisonous, corrosive, flammable gas that can be released due to the drilling or production process. As the drill bit passes through different layers in the subsurface, the gas can escape and rise to the surface through the well. At low concentrations, it smells like rotten eggs but, shortly after, temporarily eliminates the sense of smell. As a result, personnel in the area may not be aware of the presence of this gas, which has the capability to kill in a very short space of time. Hence, if hydrogen sulfide is expected in the area, detectors are positioned around the oil rig to ensure that any release cannot reach the workers undetected and, if present, individuals can vacate the area quickly without any effects. I have been involved in several projects where there was a risk of this gas emerging from the well. On my single experience when the detectors activated, the area was quickly evacuated and all personnel were directed inside the accommodation block until the wind had dispersed the gas away from the platform deck.

Due to the extensive systems of pipes and equipment used to control a well, high pressure systems are common and must be treated with extreme care. As standard, all personnel operating pressurised equipment must obtain appropriate company and industry standard training prior to use. Pressurised equipment is present in many situations but arguably the most important function is to control the pressure in the well.

While drilling, it is important to monitor and adjust the weight of the mud to at least equalise against the pressure within the formation. This prevents hydrocarbons rushing into the lower pressure wellbore from a high-pressure formation, accelerating up the well and "blowing out" of the top, where the oil rig and offshore workers are situated. Once the drilling process is complete, pressure control equipment is utilised and continuously monitored to counter any sudden changes in pressure from formation collapse or from pre-planned operations (such as opening closing valves to commence well flow or punching holes in the casing as part of the procedure to initiate controlled hydrocarbon production).

*Figure 16 - Supply boat workers* [xii]

Suspended loads are a daily hazard tackled in the industry. Cranes transfer equipment between the rig and the supply boats. They also rearrange equipment on the decks throughout the day to position equipment as it will be required for upcoming operations. Therefore, it is critical that others working in this area maintain awareness of the whereabouts of the crane and what it will be shifting.

Finally, as many of the permanent installations have existed for around 50 years, a lot of the original structure has oxidised due to harsh weather. Rusty grating and pipework is a common sight on many installations I have visited and great care must be taken in these areas. Aged equipment is potentially more likely to fail and break apart or may not be able to support your weight. Unless unavoidable, it is advisable to refrain from entering such areas; report any damaged unmarked objects on the installation so they can be inspected.

## Technology

The best way for the industry to improve profits, increase operational efficiency and     safer is through the application and introduction of technology. Cutting edge technology is apparent everywhere one looks in the industry, making it an attractive career prospect for many.

*Figure 17 - Iron Roughneck [lxiii]*

Most drill floors now make use of an automated "iron roughneck". This machine takes on the task of connecting and disconnecting pieces of pipe together on the drill floor in a uniform manner during the drilling process. This used to be done manually by a drill crew, where this repetitive and hazardous task had the potential to cause injuries. This has now been eliminated with the addition of this brilliant machine.

Drones are now used to inspect the underside of installations to inspect areas for rust, damage and potential replacement. This used to be executed by a team that hung over the side on ropes. Furthermore, inspections done by drones can take a few days where it would have taken a dedicated crew a few weeks to complete the same task.

Digital cloud technology has also allowed vast amounts of company data to be stored in a central location and easily accessible (while tightly security protected) to all employees to allow analysis and decisions to be made at an accelerated rated. As a result, operating times have been reduced, therefore reducing risk through improved fatigue management and planning.

Sensor technology has been deployed recently to monitor and predict maintenance requirements. This provides significant cost savings for operating companies, where entire structures underwent maintenance based on recommended time periods exposed rather than actual measured wear or reduction in functionality.

Advanced automation and internet connections now means that increasing numbers of operations can be controlled from onshore teams rather than a crew at the wellsite. This creates cost and safety savings in transporting less personnel to the wellsite, faster response times to challenges with on hand technical support closer by and improved safety with less workers required on installations.

Furthermore, 3-dimensional, time lapse seismic reservoir models (also known as 4-D) has enabled reservoir engineers to predict fluid behaviour and thus optimise well placement and design. As a result a greater percentage of the total volume of hydrocarbons than ever before can be removed from a reservoir. This enhanced recovery rate provides increased revenues for the oil company.

## The Ocean

Once the helicopter has vacated the helideck, generally you will remain there until either your piece of work is finished, you reach the end of your working rotation spell offshore or if there is an emergency evacuation. Looking over the side, aside from the occasional flare of another distant installation, potentially some supply boats and birds is the big blue ocean, a huge expanse of water which stretches all the way to the horizon.

*Figure 19 – Marine Wildlife* [xiv]

On memorable trips in offshore West Africa my sightings have included hammer head sharks, dolphins, whales, arrays of brilliantly colourful fish and stingrays. It is worth mentioning at this point that measures are taken to ensure that the operations carried out on oil rigs take place with substantial consideration for the environment. For example, when a new well is drilled, to obtain data regarding the formation below the surface, a borehole seismic survey can be performed. This involves sending loud powerful sonic waves from the ocean's surface down to the sea bed and into the formation. The returning sonic pulses from each layer of the formation allows a variety of properties to be determined. However, independent research argues that powerful sonic waves can disrupt the navigational abilities of ocean animals such as whales and dolphins. Hence, during such operations, an environmental representative is present the installation and surveys the ocean. If any animals that may be affected by the seismic operation are spotted, then the operation can be paused until it has passed beyond a predetermined distance before resuming.

Returning North to cooler waters, beagle sharks have been spotted in the North Sea on occasion. However, to one colleague's surprise, they were glancing over the side when a submarine surfaced and passed by. Presumably this must have been either the British or Norwegian military during one of their ventures. Clearly it wasn't deterred by the possibility of witnesses reporting its whereabouts.

Seagulls must receive a special mention and paragraph here. These loud squawking birds venture hundreds of miles out to sea by nesting on the supply boats as they travel continuously between the oil rig and the shore. One must admire their resolve to travel out this distance to sea, particularly when the vessel they follow and nest on is not used for fishing. Despite this, it is rare to lay eyes on a malnourished seagull. All tend to appear well fed, clean and proud; with an eye to gobble any scrap of food found abandoned.

The industry in general has a mixed reputation with the rest of the world, particularly when considering its environmental impact. From personal experience, the industry does make significant effort to minimise its environmental impact when it is economically feasible to do so. Companies continually work to balance the demands and costs of producing hydrocarbons while minimising the impact on the environment. Specific charities and groups have been known to demonstrate against the operations of companies in the industry and my sole experience of this presented itself in the form of a well-known group renowned for working to defend the natural world. A small boat approached the side of the oil and gas platform and most of the individuals embarked to ascend the oil rig via ropes and ladders that were stationed over the side (primarily used as a means of escape if necessary). It took them some time to scale the platform and some tired and fell back into the ocean. One man was so exhausted from his efforts to scale the platform legs to the main deck that he was taken directly to the medic by two of the on-shift workers before being provided with some dry clothes and then guided to the galley to warm up. All protestors were taken on board and kept comfortable before a helicopter arrived to return them to dry land.

In stretches of water, where the distance between two pieces of land may be overcome by semi-amateur sea adventurers, civilian boats can also be identified. I recently learned of an annual boat race where competitors embark one year from Norway and the other Scotland to reach the other nation's shores. This usually means passing by many oil rigs during the journey. Viewing this spectacle from the platform itself is brilliant, as tall white and classic sailing boats pass by. It makes a welcome change from the normal view of supply boats.

Finally, in the far North, if the conditions are right and one happens to be glancing in the right place at the right time, the Aurora Borealis can be enjoyed from the deck of an oil and gas platform. Tourists pay top dollar for excursions to the Arctic circle in the hope of viewing this phenomenon. Here, if one has luck on their side, it can be viewed for free! The near zero artificial lighting from cities and road lights can make clear viewing possible.

## New Recruits

The energy industry is arguably the world's most important. Currently, and for the foreseeable future, oil and gas exploration and production will be the greatest contributor for meeting global energy demand and as a result the industry has a bright future. Furthermore, as renewable energy gains momentum and begin to replace oil and gas, companies in the industry will adapt, rather than disappear, to facilitate the growth and success of these energy sources. Therefore, I would encourage anyone to get involved in this industry.

Graduates & new hires can expect to receive higher than average salaries (average UK starting salaries in 2017 was 28,000 GBP before bonuses or overtime) and, based on your performance and initiative, salaries can become quite lucrative. Due to the well-publicised skills shortage, employees can expect to be given real responsibility and accountability early on in their careers and develop many skills such as managing teams, organising millions of dollars' worth of assets and co-operating as part of a team of diverse nationalities, backgrounds and skills sets.

Female employees are highly sought after. The industry continues to diversify and needs many more women in this currently male-dominated industry. Executive management teams have placed gender equality at the top of their agenda for improving their company's performance. Similarly, to multicultural teams, multi-gender teams enhance department performance and output. Various internal women's groups have also been set up so that women can collaborate with full confidence for overcoming potentially challenging situations at work.

With so many different roles within companies and various companies involved with oil and gas, a borderless career opportunity awaits those joining the industry. One might find themselves at the wellsite in an operations capacity, onshore in a technical support role, possibly in sales and marketing, as a lawyer, software and information technology, human resources, logistics coordinator or design engineer. There are many other in-demand roles required to support the industry as well. Careers are not restricted in terms of location, with opportunities to work on every continent. It is very advantageous if employees speak more than one language (with the ability to speak, read and write English as a minimum).

Finally, as the remaining oil and gas fields continue to be housed in more challenging and harsher environments, cutting edge technology, must be used

challenging and harsher environments, cutting edge technology, must be used

**Conclusion**

From the very first well drilled over 100 years ago, oil and gas has developed into the largest business in the world, meeting global energy demand.

To meet this, people must go to the farthest ends of the earth to find and produce it. Often this means working in the ocean. Working on an offshore installation brings many challenges and exciting opportunities; with colleagues collaborating both on land and at the rig site to accomplish tremendous tasks. The makes this career particularly rewarding to those who take an interest in engineering, safety and world progress.

Individuals can expect to find themselves working with advanced technology, rivalling that used in the jet plane or space industries. They can also expect fast paced, global careers with significant responsibility and accountability and a huge backdrop of support from highly skilled teams around the world.

As the global population increases and developing countries continue to **modernise**, the demand for energy will also inflate. The oil and gas industry will, therefore, continue to be the provider of modern **civilisation's** energy needs for the foreseeable future.